ANA
THE

THINGS YOU SHOULD KNOW
(QUESTIONS AND ANSWERS)

By Rumi Michael Leigh

Introduction

I would like to thank and congratulate you for purchasing this book, *"Anatomy and physiology, the reproductive system, things you should know (questions and answers)"* series.

This book will help you understand, revise and have a good general knowledge and keywords of the human anatomy and physiology.

Thanks again for downloading this book, I hope you enjoy it!

Questions: Part 1

1. What are the phases of the ovarian cycle?
2. How long does the follicular phase last?
3. How long does the luteal phase last?
4. What is spermatogenesis?
5. How long does the spermatogenesis cycle last?
6. There are how many phases of the spermatogenesis cycle?
7. What are the phases of the spermatogenesis cycle?
8. What is the duration of mitosis of spermatogonia?
9. What is the duration of meiosis 1?
10. What is the duration of meiosis 2?

Answers: Part 1

1. The follicular phase and the luteal phase.
2. It lasts 14 days. The first 14 days.
3. It lasts 14 days. The fifteenth day to the twenty-eighth day.
4. Spermatogenesis is the formation of spermatozoa.
5. The cycle lasts 64 days.
6. There are 4 phases.
7.
- Mitosis of spermatogonia.
- Meiosis 1.
- Meiosis 2.
- Spermiogenesis.
8. 16 days.
9. 24 days.
10. A few hours.

Questions: Part 2

1. In the phase of spermatogenesis, what happens during the mitosis of spermatogonia?
2. In the phase of spermatogenesis, what happens during meiosis 1?
3. In the phase of spermatogenesis, what happens during meiosis 2?
4. In the phase of spermatogenesis, what happens during spermiogenesis?
5. What is a condensed DNA molecule?
6. What does DNA contain?
7. What are the 2 cell phases?
8. Explain mitosis.
9. Explain interphase.
10. Which of the 2 cell phases is the longest?

Answers: Part 2

1. The formation of primary spermatocytes.
2. The division of primary spermatocytes into secondary spermatocytes.
3. The formation of spermatids.
4. The maturation of spermatids.
5. A chromosome.
6. DNA contains genetic information.
7. Mitosis and interphase.
8. Mitosis is cell division.
9. Interphase is cellular replication.
10. The interphase.

Questions: Part 3

1. What is the duration of mitosis?
2. What is the duration of the interphase?
3. How many pairs of chromosomes do we have?
4. What are the 4 steps of mitosis?
5. What is a single chromosome?
6. What is a double chromosome?
7. What are the annexes genitals?
8. What is the muscle of the uterus?
9. What is oogenesis?
10. What are the phases of the ovarian cycle?
11. Is the nuclear maturation phase a complete phase?

Answers: Part 3

1. Approximately 40 minutes.
2. About 24 hours.
3. We have 23 pairs of chromosomes.
4. Prophase, metaphase, anaphase and telophase.
5. A single chromosome consists of a single chromatid.
6. A double chromosome consists of two chromatids.
7. Glands and external genitals and ducts.
8. The myometrium.
9. Oogenesis is the formation of the egg in women.
10. A phase of multiplication, a phase of nuclear maturation and a cytoplasmic phase.
11. No, it is not complete.

Questions: Part 4

1. What irrigates the testicles?
2. What drains the testicles?
3. What is the perineum area?
4. What is prostatitis?
5. What is the circumcised part of a man?
6. Is sperm alkaline or acidic?
7. What is the pH of sperm?
8. What is the use of sperm coagulation?
9. Name a few sexually transmitted infections.
10. Are herpes bacterial infections or virus infections?

Answers: Part 4

1. The testicular arteries.
2. The testicular veins.
3. It means around the anus.
4. Prostatitis is the inflammation of the prostate.
5. The foreskin.
6. It is alkaline.
7. From 7,2 to 8.
8. Sperm coagulation allows the sperm to remain glued to the walls of the vagina.
9. AIDS, syphilis, herpes, gonorrhea, etc.
10. Herpes are bacterial infections.

Questions: Part 5

1. When does menopause begin?
2. What happens to the mobility of spermatozoa with age?
3. What governs the course of puberty?
4. Which muscles prevent the evacuation of urine during ejaculation?
5. Is ejaculation governed by the sympathetic or parasympathetic system?
6. Is erection governed by the sympathetic or parasympathetic system?
7. What stimulates oogenesis?
8. What marks the beginning of a new menstrual cycle?
9. What is orchitis?
10. What is dysmenorrhea?

Answers: Part 5

1. From the age of 45.
2. The speed of its mobility is reduced.
3. The hypothalamus.
4. The internal urethral muscles.
5. The sympathetic system.
6. The parasympathetic system.
7. Estrogens.
8. Menstruation.
9. It is the inflammation of the testicle.
10. It is painful menstruation.

Questions: Part 6

1. What is the function of the scrotum muscles?
2. In relation to question 1, how?
3. What is amenorrhea?
4. What is the ligament that supports the uterus?
5. What is the hormone that triggers ovulation?
6. What is the name of the embryo after 8 weeks?
7. What is the function of oxytocin?
8. What are the male gonads?
9. Which organ forms the spermatozoa?
10. What is the function of the prostate?

Answers: Part 6

1. They help us maintain the temperature of the scrotum.
2. When the temperature is low, the scrotum muscles contract to seek body heat and when the temperature is high, the scrotum muscles relax in order to reduce its temperature.
3. Amenorrhea is the absence of menstruation.
4. The round ligament.
5. The LH hormone (luteinizing hormone).
6. The fetus.
7. It is a hormone that causes milk ejection in a woman.
8. The testicules.
9. The testicules.
10. It secretes a fluid that activates the spermatozoa.

Questions: Part 7

1. What happens when the secretion of progesterone stops?
2. What are the male gametes?
3. Is the pH of the vagina alkaline or acidic?
4. What are the female gonads?
5. What are the female gametes?
6. What is the maximum lifespan of the egg?
7. The ovary usually releases how many eggs at a time?
8. What is the organ of gestation?
9. What is an oocyte?
10. Where does fertilization take place?
11. How long does the ovarian cycle last?

Answers: Part 7

1. Menstrual flow begins.
2. The spermatozoa.
3. It is acidic.
4. The ovaries.
5. The ova.
6. One day.
7. Only one egg at a time.
8. The uterus.
9. The oocyte is an immature egg.
10. Fertilization takes place in the fallopian tubes.
11. It lasts 28 days.

Questions: Part 8

1. What is the skin that covers the glans?
2. What is the luteal phase?
3. What is the ovarian follicle?
4. What is the maximum life span of the spermatozoon?
5. What is Bartholin?
6. Which gland neutralizes the acidity of the urine?
7. What is human reproduction?
8. Which of the testicles is higher?
9. How many chromosomes are there in a spermatozoon?
10. What is the function of the epididymis?
11. What is the function of the testicles?

Answers: Part 8

1. The foreskin.
2. This is the third phase of the menstrual cycle.
3. It is a set of cells and oocyte.
4. About 4 days.
5. It is the gland that lubricates the vagina.
6. The Cowper.
7. This is the process that allows two individuals (male and female) to increase their numbers by giving birth to a new individual.
8. The right testicle.
9. 23 chromosomes.
10. The storage and maturation of spermatozoa.
11. The manufacture of spermatozoa and testosterone.

Questions: Part 9

1. How does the testicles produce spermatozoa?
2. What is the function of the vagina?
3. The urethra is part of which system?
4. What is the function of the uterus?
5. What is the function of the ovaries?
6. What is the function of the vas deferens?
7. What is the function of the penis?
8. What is the function of the seminal vesicles?
9. Is the liquid produced by the seminal vesicles acidic or alkaline?
10. What is the function of the prostate?

Answers: Part 9

1. They make them by meiosis.
2. It is the female's copulatory organ.
3. The urethra is part of the reproductive system and the urinary system.
4. It is a muscular organ that helps during childbirth and menstruation.
5. The production of ova and hormones (estrogen and progesterone).
6. The vas deferens transports the spermatozoa.
7. It is the copulatory organ of a man.
8. It produces a liquid that nourishes and activates the spermatozoa.
9. It is alkaline.
10. The activation of spermatozoa by the secretion of a liquid.

Questions: Part 10

1. How many oocytes are there at birth?
2. How many oocytes are there at puberty?
3. Which cells secrete testosterone?
4. What is menopause?
5. What is in the head of a spermatozoon?
6. What produces the energy in a spermatozoon?
7. What part of the spermatozoon contains enzymes?
8. What part of the spermatozoon allows its movement?
9. Where do female sex cells develop?
10. Where are the female sex cells released?

Answers: Part 10

1. There are about 500,000 oocytes per ovary.
2. There are about 200,000 oocytes per ovary.
3. Interstitial cells.
4. Menopause is a stage when a woman begins to lose her reproductive capacity.
5. The nucleus (it contains DNA).
6. Mitochondria.
7. The acrosome.
8. Its tail.
9. They develop in the ovaries.
10. They are released into the uterus.

Questions: Part 11

1. What are some of the reasons for circumcision?
2. What is detumescence?
3. What are some factors related to erectile impotence?
4. What is the ideal temperature for sperm formation?
5. Why are male gonads mixed glands?
6. What are some endocrine functions of testosterone?
7. What is responsible for the endocrine regulation of the sexual function?
8. What is the weight of the prostate?
9. Is ejaculation voluntary or involuntary?
10. Give an example of an involuntary ejaculation.

Answers: Part 11

1. Religious, hygienic and medical reasons.
2. Detumescence is the end of erection.
3. Taking some medications, alcoholism, diabetes, neurological diseases, etc.
4. About 35 degrees Celsius.
5. Male gonads are mixed glands because they have endocrine functions and exocrine functions.
6.
- The stimulation of spermatogenesis.
- The masculinization of the brain.
- Erection.
- Ejaculation.
- Growth of the hairs (the beard, etc.).
- Increased bone mass.
- Lower voice.
- Acne.
7. The hypothalamus, the hypophysis and the testicles.
8. About 20 g.

9. Ejaculation may be voluntary or involuntary.
10. Ejaculation that occurs during sleep.

Questions: Part 12

1. What happens to the pH of the vagina after menopause?
2. What are some of the consequences of menopause?
3. During erection, the penile arteries become?
4. What happens to the penile veins during erection?
5. What are the erectile tissues of the penis?
6. How many chromosomes have a second-order oocyte?
7. What do the high estrogen and progesterone levels cause during pregnancy?
8. The decline in testosterone begins after what age?

Answers: Part 12

1. The pH becomes alkaline.
2.
- Weight gain.
- The dryness of the skin.
- Vaginal dryness.
- Osteoporosis.
- Hair loss, etc.
3. Dilated.
4. They compress.
5. The muscular bodies and the cavernous bodies.
6. 23 chromosomes.
7. High levels of estrogen and progesterone cause thickening of the uterine lining.
8. After the age of 50.

Conclusion

Thank you once again for purchasing this book. I hope it has helped you in your journey to understanding the anatomy and physiology of the human body.

Please, if you enjoyed this book, I would like you to leave a review. It'd be appreciated.

Thank you.

CPSIA information can be obtained
at www.ICGtesting.com
Printed in the USA
LVHW100306260722
724422LV00004B/105